Materials

Kay Davies
and
Wendy Oldfield

Starting Science

Books in the series

Animals
Day and Night
Electricity and Magnetism
Floating and Sinking
Food
Hot and Cold
Information Technology
Light
Local Ecology

Materials
Plants
Pushing and Pulling
The Senses
Skeletons and Movement
Sound and Music
Waste
Water
Weather

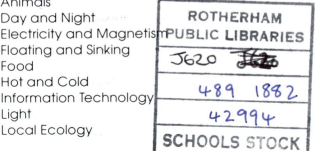

About this book

This book on materials introduces children to the many useful and decorative treasures our earth holds. **Materials** includes information and activities on a selection of gases, solids and liquids. Children learn about the properties of natural and synthetic materials, how we use them and the importance of conserving them for the future.

This book provides an introduction to methods in scientific enquiry and recording. The activities and investigations are designed to be straightforward but fun, and flexible according to the abilities of the children.

The main picture and its commentary may be taken as an introduction to the topic or as a focal point for further discussion. Each chapter can form a basis for extended topic work.

Teachers will find that in using this book, they are reinforcing the other core subjects of language and mathematics. Through its topic approach **Materials** covers aspects of the National Science Curriculum for key stage 1 (levels 1 to 3), for the following Attainment Targets: Exploration of science (AT 1), Human influences on the Earth (AT 5) and Types and uses of materials (AT 6).

First published in 1991 by
Wayland (Publishers) Ltd
61 Western Road, Hove
East Sussex, BN3 1JD, England

© Copyright 1991 Wayland (Publishers) Ltd

Typeset by Kalligraphic Design Ltd, Surrey
Printed in Italy by
  Rotolito Lombarda S.p.A., Milan
Bound in Belgium by Casterman S.A.

**British Library Cataloguing in
  Publication Data**
  Davies, Kay
  Materials. – (Starting science)
  I. Title II. Oldfield, Wendy III. Series
  600

ISBN 0 7502 0207 6

Editor: Cally Chambers

CONTENTS

All the words that first appear in **bold** in the text are explained in the glossary.

OPEN THE BOX

Our earth is like a box of treasures. Everywhere we look we can find beautiful and useful materials.

Cotton T-shirt

Woollen jumper

Wooden spoon

Metal key

Coal fire

China cup

Can of oil

Glass bottle

Plastic tub

Animals' hair and parts of plants are used for clothes. Trees give us wood for our houses. Over millions of years, fallen trees can become shining black coal deep in the earth.

We also use materials that have never been alive.

Layers of rocks and metals under the ground can be used for roads, bridges and buildings.

Which of these materials have been alive? Which have never been alive? Use this book to help you find out.

Object	Once alive	Never alive
Cotton T-shirt		
Wooden spoon		
Metal key		

Can you add some materials of your own to this list?

Birthday presents are fun to open. Everything inside the
party box is made from some of the earth's treasures.

Air is all around us. It is so light we hardly notice it, but it holds the hang-glider high above the ground.

AS LIGHT AS AIR

Air is a mixture of **gases**. It is always pushing in all directions around us, but we only feel it on a windy day.

Air can be a useful tool.

Using a pump, we squash air into tyres for bikes, cars and lorries.

The squashed air in the tyres is like a cushion. It gives us a smooth ride.

Blow up a balloon and tie the neck.

Cut a wheel as round as your balloon from a stiff piece of card. Push a pencil through the centre.

First roll your wheel, then the balloon, over a bumpy surface.

Can you feel a difference?

Trees are sawn into long pieces of wood. The wood is sent to **factories** to be made into many useful things.

WOOD WORK

Wood comes from trees. We can find lots of uses for it. How many uses do you recognize?

When we cut down trees we can plant new ones. Then we need never run out of wood.

Balsa wood is very soft and light. Ask an adult to cut a piece into a rough boat shape.

Now use sandpaper to finish shaping your boat.

Make a mast and sail out of a straw and paper. Stick it on to the boat with some clay.

Does your boat sail?

ALL WRAPPED UP

Paper and card are made from wood. We use them to write and print on. We use them to wrap parcels.

Are there lots of other ways of using paper?

Here's a puppet for you to make. It is easy to make shapes out of paper.

Cut out a square piece of card. Roll it into a tube shape and stick the edge down.

Cut slits in a long, narrow piece of paper to make a comb shape.

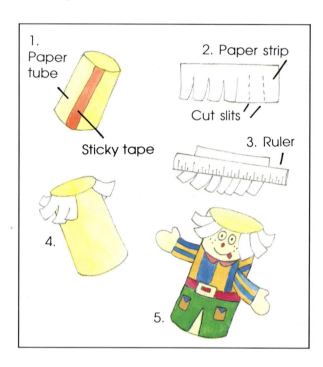

1. Paper tube

Sticky tape

2. Paper strip

Cut slits

3. Ruler

4.

5.

Rub a ruler hard down the strips to stretch and curl the paper.

Stick your curly paper round the top of the tube.

Paint a face and clothes on your puppet. You can add paper arms too.

Paper has been cut and folded to make these models.

STRINGY THINGS

String is made from thin strips of plants.
Each strip is called a **fibre**. String can be made from thin plastic fibres too.

Lots of fibres can be twisted together to make rope.
When fibres are twisted together they are very strong.

Try to count the fibres at the end of pieces of string.
Does thick string have more fibres than thin string?

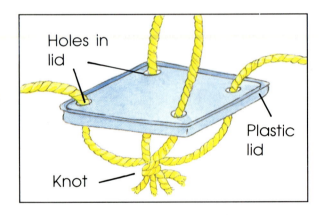

Holes in lid

Plastic lid

Knot

Find four pieces of string about 70 cm long.

Tie them together with a knot at the bottom.

Thread them through holes in a large plastic lid.

Tie the string at the top.

Place a flower pot on the lid to make a hanging basket.

The string is **dyed** with colour at the factory.
Machines wind it into giant reels.

The wool on sheep's backs keeps them warm.
It can be spun and woven to make warm clothes for us.

WOOLLY JUMPER

Fibres come from plants and from animals too. These are called natural fibres. The hairs on your head are natural fibres.

Some fibres are made from plastic. We call these **synthetic**.

We knit and **weave** fibres to make cloth for clothes. Cloth has tiny holes in it. These holes trap warm air to keep out the cold.

Look at the labels in your clothes.
Make a list of what they are made from. Are they natural or synthetic? Do the synthetic fibres look and feel like the natural ones?

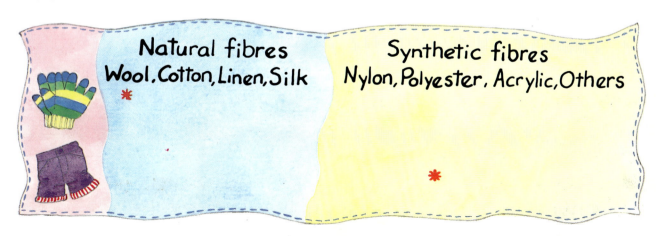

Natural fibres
Wool, Cotton, Linen, Silk

Synthetic fibres
Nylon, Polyester, Acrylic, Others

AS OLD AS THE HILLS

Rocks and stones are found in the ground.

There are many kinds. Some are useful and some are beautiful.

Some have taken millions of years to be made.

Miners dig rocks and stones out of the ground.

Make a stone collection.

Try to find some broken stones. They may have coloured layers inside.

Paint them with varnish to make them shine.

Show them off in a glass jar or use one for a paperweight.

The carvings on the pillar have been cut into hard stone. They will last for many years.

The clay is mixed well and **moulded** into bricks.
They will be used to build walls and houses.

MODEL MATERIAL

We find clay in the earth. When it is wet it can be shaped into bricks, tiles and **pottery**. If it is left to dry and heated in a **kiln**, it becomes very hard.

You can use wet clay to make a pot like this:

1. Roll a long clay snake between your hands.

2. Cut out a base for your pot.

3. Wind the clay around the base.

4. Run your thumbs down the sides to join the coils together.

Leave your pot to dry, then paint it.

A coat of varnish will make your pot **waterproof**.

What other kinds of pot could you make using a coil of clay?

The glass in the windows is **transparent**. It lets light into the rooms and lets people look out at the world.

CLEAR AS GLASS

Glass is made from materials found in sand.

When these materials are heated they become soft and stick together. The red-hot glass can be stretched, bent and twisted to make any shape we want.

Glass becomes smooth and **rigid** when it cools.

Collect empty glass bottles. Find different shapes and colours.

Drop some coloured beads inside. Are all the bottles transparent? What do the beads look like?

When you have finished with your bottles, you can take them for **recycling**. They will be melted down and made into new glass.

HARD AS NAILS

There are many sorts of metals. They can be very different; some hard, some soft and some shiny.

Metals are used for all sorts of things.
Coins are made from metals by cutting out round shapes and pressing patterns on to both sides.

Some metals can be beaten into flat sheets. Some can be pulled into thin wires to carry electricity.

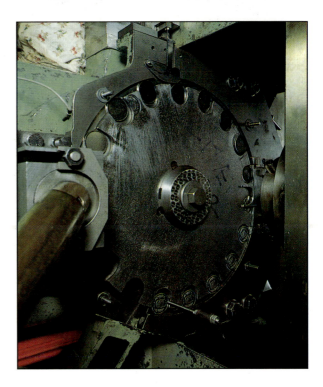

Why don't you sort some metals for recycling?

Start collecting empty drinks cans. Some will be made from aluminium and some from steel.

Use a magnet to test them. Only steel cans will stick to the magnet.

We can melt some metals by heating them. They can be poured into moulds to make the shapes we want.

STIR FRY

We use oil from plants and fish for cooking and eating.

Some oil is found under the ground. This comes from the remains of plants and animals buried millions of years ago. We can't eat it. It makes petrol, and oil for heating buildings.

It can be used to make plastics, paints and medicines.

Colour some water in a jar with food colouring or paint.

Pour some oil into the water. Put the lid on and shake it hard. Leave it.

What happens to the oil? Does it mix with the water?

We can use oil to protect things from water.

The Chinese food is fried in oil heated in the pan.
Oil stops the food sticking and helps it cook.

Rubber is made from the sticky **sap** of the rubber tree.

BOUNCING BACK

Rubber can be made soft and stretchy for elastic bands or hard and tough for tyres.

Rubber can change shape easily. This makes it bouncy.

We can play lots of games with rubber balls.

Find all sorts of rubber balls.

Stand a metre rule on the floor.
Drop each ball in turn from the top of the rule.

Ask a friend to record how high each ball bounces.

Make your results into a chart to show how bouncy the balls were.

The giant model is made out of hollow plastic bricks.
They can be pressed together and pulled apart.

PLASTIC FANTASTIC

Plastic is made mostly from oil or coal. It is synthetic.

Sheets of new plastic are warmed until they bend.

They can be moulded to any shape.
When plastic cools it becomes hard and waterproof. It can be made into a rigid bowl or bendy kitchen film.

Make these *Jelly Jollies* using a moulded plastic tray from a chocolate box.

Make up some jelly. Use less water than it says on the packet. Pour it into the moulds and let it set.

Dip the plastic base in warm water for a few seconds.
Tip your *Jelly Jollies* on to a plate. Find a plastic spoon to eat them with.

GLOSSARY

Dyed Made a different colour.

Factories Places where machines are used to make things.

Fibre A fine thread made from a natural or synthetic material.

Gases Materials, like air, which are neither liquid nor solid.

Kiln An oven used for baking clay.

Miners People who dig for rocks and metals underground.

Moulded Made into shapes.

Pottery Containers made from clay.

Recycling Specially treating rubbish so that it can be used again.

Rigid Hard and firm.

Sap The juice from inside a tree.

Synthetic Materials that are not natural.

Transparent See-through.

Waterproof When water can't pass through a material.

Weave To thread spun fibres in and out to make cloth.

FINDING OUT MORE

Books to read:

Air by Angela Webb (Franklin Watts, 1989)
Bangles, Badges and Beads by Chris Deshpande (A & C Black, 1990)
Materials by Terry Jennings (OUP, 1984)
My Jumper by Robert Pressling (A & C Black, 1989)
Rock Collecting by Roma Gans (A & C Black, 1987)
Waste by Kay Davies and Wendy Oldfield (Wayland, 1990)

You may also find some useful books to read in the following series:

Links (Wayland)
Material World (Wayland)
Our Clothes (Wayland)
Threads (A & C Black)

Teachers' Packs:

Materials, Me & My Senses, and **Materials** from Scholastic Publications, Westfield Road, Southam, Leamington Spa, Warwickshire CV33 0JH.
Materials from Molehill Press, Grange Farmhouse, Geddington, Kettering, Northamptonshire NN14 1AL.

PICTURE ACKNOWLEDGEMENTS

Cephas 18, 25; Eye Ubiquitous 20; Geoscience Features 16 top; J Allan Cash Ltd. 26; Edward Parker 15; Tarquin Publications 11; Tony Stone Worldwide 6, 7 top, 14, 23, 28; Wayland Picture Library (Zul Mukhida) *cover*, 5, 7 bottom, 9, 10, 12, 16 bottom, 19, 21 bottom, 22 bottom, 24 bottom, 27 bottom, 29; ZEFA 8, 17, 21 top, 22 top, 27 top.
Artwork illustrations by Rebecca Archer.
The publishers would also like to thank the teachers, parents and children of Davigdor Infant's School of Hove, East Sussex, for their kind co-operation.

INDEX

Page numbers in **bold** indicate subjects shown in pictures, but not mentioned in the text on those pages.